Animals Around Us

Animals of the Woods and Forests

by Julie Becker

EMC Publishing, St. Paul, Minnesota

**for
all the animals
who live
in the green mountains
of Vermont**

Library of Congress Cataloging in Publication Data

Becker, Julie
  Animals of the woods and forests.

  Includes Index
  (Her Animals around us)
  SUMMARY: Text and illustrations introduce ten animals
living in the woods. Included are the skunk, walking-
stick, red fox, owl, porcupine, chipmunk, woodpecker,
opossum, red newt, and little brown bat.
  Forest fauna—Juvenile literature. [1. Forest
animals] I. Roth-Evenson, Maarja. II. Title.
III. Series.
QL112.B38   591.9'09'52   77-8253
ISBN 0-88436-396-1

Copyright 1977, 1982 by EMC Corporation
All rights reserved. Published 1977.
Revised Edition 1982.

Published by EMC Publishing
180 East Sixth Street
St. Paul, Minnesota 55101
Printed in the United States of America
0 9 8 7 6 5 4 3 2

TABLE OF CONTENTS

Woods and Forests

Woods and forests are full of trees. Leaves and needles grow on the trees. Small plants and flowers grow under the trees. Bright green moss grows on the ground.

Woods are smaller than forests. Woods are near open spaces. Many animals live in the woods and forests. Some live in the trees. Some live under rocks and logs. And some live under the ground.

The animals in the woods and forests are hard to find. Many of them hide in the daytime. They hide when people come. But they are there.

Let's follow a path. Let's walk among the trees. Let's find some animals that live in the woods and forests.

The Skunk

This skunk walks in the forest very slowly. He is not in a hurry. He is a little animal. But he is not afraid of big animals. Most big animals stay away from the skunk.

Other small animals in the forest run very fast. They hide when they see big animals. But not the skunk. He won't hide. He knows no one will hurt him.

The skunk has his own spray gun. His spray gun is under his tail. He can shoot out a watery spray. This spray has a yellow color. It has a very bad smell.

The skunk's spray is very strong. He can shoot his spray ten or fifteen feet. But the smell of his spray can go much farther. The bad smell can go a mile away.

If the skunk is going to shoot his spray, he will turn around. He will pound his front feet on the ground. He will raise his back. He will stick his bushy tail straight up into the air.

When the skunk shoots his spray, he has good aim. He will not miss. If a dog tries to hurt a skunk, he will be sorry. The skunk will spray him right in the face.

The skunk's spray may go into the dog's eyes. The spray will hurt the dog's eyes. It will make them burn. But the dog will never go near a skunk again!

The skunk will not spray at every animal. He only sprays when he is in danger. He is not glad to shoot his spray. He just wants to be left alone.

The skunk hunts
for food in the night.
He digs in the
ground. He digs with
his sharp claws. He
hunts for worms and
insects. He finds
them in the ground.

The skunk eats
many kinds of food.
He eats mice. He eats
small birds. He eats
eggs and fruit and
fresh green plants.

When the night is over, the skunk walks home. His home is a hole under the ground. He may live in a hole that another animal made a long time ago.

When he gets home, he washes himself. He combs his fur with his sharp front claws. He licks his paws so he can wash his face. When he is clean, he will go to sleep.

The skunk is fun to watch. He is a friendly little animal. But he will spray if he thinks he is in danger. If you see a skunk in the woods, don't be afraid. Move slowly. Talk in a soft quiet way. But to be on the safe side, don't get too close!

The Walkingstick

There is an insect in the woods that looks just like a stick. She looks like a stick that can walk. She is a walkingstick.

The walkingstick is about as long as your finger. She is about as thick as a flower stem. She is brown.

The walkingstick comes out of a little egg. She comes out in the spring. When she first comes out, she is green. The new twigs on the trees are green too.

When the new twigs get bigger, they turn brown. The walkingstick turns brown too. She turns the same color as the twigs. If she is the same color, she will be safer. She will be hard to see.

The walkingstick lives in a tree.
She eats leaves. She bites holes in
the middle of the leaves. She nibbles
on the sides of the leaves. She likes
to eat oak leaves best of all.

Many insects can fly. But the
walkingstick cannot fly. She has no
wings. She stays in her tree and
eats the leaves. When she isn't
eating, she stands very very still.
She hopes the birds will not see her.
If they see her, they will eat her.

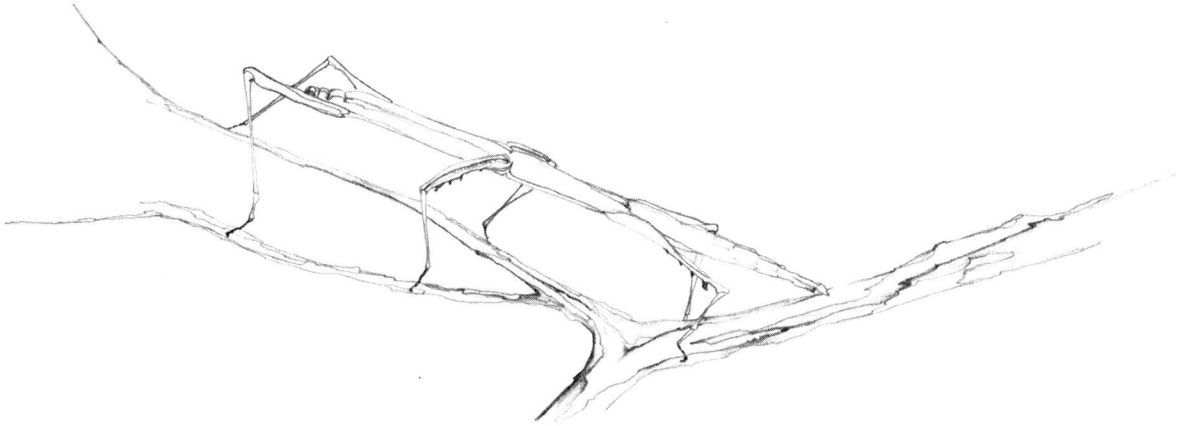

The walkingstick has six feet. She has little hooks at the end of each foot. She holds on to a twig with her six little hooks. She holds on tight. She can stand in one place for hours without moving.

In the fall, the walkingstick lays her eggs. She stands on a branch. The eggs drop out of her body one by one. Over a hundred eggs drop out. They all fall to the ground. The eggs are black and white. They are round. They look like tiny little peas.

The wind blows. Leaves fall from the trees. The leaves cover up the walkingstick's eggs. The eggs stay under the leaves all winter long. It will be hard for a mouse or a bird to find the eggs.

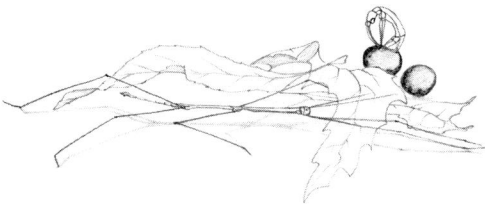

Some of the eggs will get eaten. But not all of them. In the spring, some of the eggs will hatch. There will be new walkingsticks in the woods. Each new walkingstick will climb up a tree. Each new walkingstick will begin to eat leaves.

If you walk in the woods, don't be too surprised. You may find a stick that has legs! You may find a stick that can walk!

The Red Fox

The red fox runs in the woods. He is holding a rabbit in his mouth. He will take the rabbit home. He will feed the rabbit to his children.

His children live in a den. Their den is a hole under the ground. They come out of their den to play. They come out of their den to eat too.

When the fox children were very little, they drank milk from their mother. But now they are bigger. Now they eat meat. Their father hunts for them. He brings food to them. Their mother hunts too.

The mother and the father fox both take care of their children. If the father is out hunting, the mother babysits. If the mother is out hunting, the father babysits. They take turns.

In the summer, the red fox lives with his family. He lives in the woods. His den is near an open place. It is near the fields. Sometimes the red fox hunts in the fields. Sometimes he hunts in the woods.

The red fox can move very quietly. If he sees a mouse in the fields, he follows it. Then he jumps on the mouse. He jumps very quickly. He eats the mouse for dinner.

The fox eats a lot of mice. He eats birds and rabbits too. He even eats grasshoppers. But meat isn't his only food. He eats fruit too. He loves to nibble on berries that he finds in the woods.

The red fox is a very fast runner. Sometimes dogs try to catch him. They chase him in the fields and in the woods. They can smell his tracks. They can tell where he runs.

The fox tries to fool the dogs. He plays tricks on them. He runs in circles. He runs in zigzags. He swims in the brook. He jumps across the brook. He runs along the top of a fence. Sometimes he even rides on the back of a sheep. Most of the time he gets away.

The red fox has a thick coat of fur. He has a thick bushy tail. He sleeps outside in the winter. He curls up on the ground. He curls his thick bushy tail around his face. His bushy tail keeps his nose warm.

The red fox hunts in the winter too. He finds mice running under the snow. He chases rabbits on top of the snow. When he gets thirsty, he eats snow.

When the red fox runs in the snow, his feet sink in. His feet make small round tracks in the snow. Look for his plain round tracks in the winter.

The Owl

The owl has wonderful eyes. She can see a mouse running on the ground when it is dark outside. Her eyes are like head-lights on a car. They look straight ahead. They are on the front of her face.

The owl has rings around her eyes. Her eyes are big and round and yellow. She looks like a person wearing glasses.

Some owls have short feathers on top of their heads. These short feathers stick up. They look like little horns. But they are not horns. They look like ears. But they are not ears.

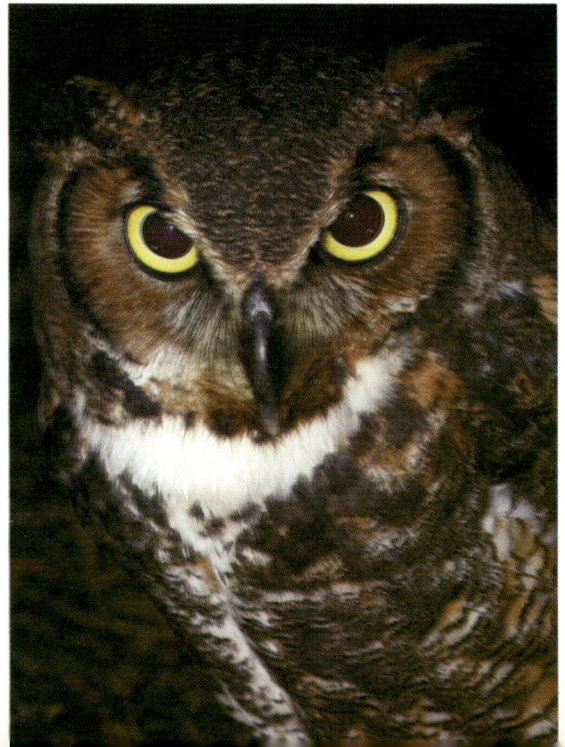

It is night. The moon shines over the forest. The owl flies in and out of the trees. She flies silently. Her wings move. But they do not make a sound.

The owl sees a mouse on the forest floor. The mouse can not see the owl. She can not hear the owl. The owl flies down. She grabs the mouse. She eats the mouse for supper.

The owl hunts at night. The moon and the stars help her to find her way. Her bright yellow eyes help her too.

The owl's ears are on the side of her face. They are covered with feathers. Most birds have small holes for ears. But the owl has little ear flaps. She is the only bird that has flaps for ears. Her ear flaps catch sounds in the air.

The owl has wonderful hearing. She can hear a mouse crawling under a plant. Her ear flaps face the ground when she flies. Her eyes face the ground too. Her eyes and ears make her a very good hunter.

Her feathers make her a good hunter too. The feathers on the owl's wings are very soft. They are fluffy. When the owl moves her soft fluffy wings, you can not hear them. She can not fly fast. But she can fly silently. Little animals in the forest can not hear her. She can surprise them.

The owl eats meat. Big owls like to eat rabbits and squirrels. Small owls like to eat frogs and mice and birds.

When the owl sees a mouse in the forest, she flies down. She grabs the mouse with her feet. Her feet are strong. The claws on her feet are sharp. The mouse can not get away.

After the owl grabs the mouse, she kills it. She kills it with her sharp beak. Then she eats it.

Many times, the owl swallows her food whole. She swallows small animals whole. She swallows small birds whole. She swallows the fur and the feathers and the bones.

The fur and the feathers help the owl. They brush out her stomach. They clean her stomach. But they do not stay in her stomach. In a few hours, the owl spits them up.

Each time the owl eats, she spits up a pellet. The pellet is round and hard. The pellet is full of bones. It is full of fur and feathers. If you cut open a pellet, you can tell what the owl had for dinner!

Some owls are large and some are small. Some are grey. Some are red-brown. And some are white. But all owls are good hunters. And all owls fly silently in the night.

If you go to the forest at night, you may hear scary sounds. Owls make these scary sounds late at night. You may hear a hoot or a laugh or a scream. If you hear these sounds, you know that an owl is near.

Listen. Do you hear a hoot, hoot? Don't be afraid. The owl is calling!

The Porcupine

The word porcupine means "a pig with spines." The porcupine is not really a pig. But his head, his back, his legs and his tail are all covered with spines. These spines are called quills.

The porcupine never seems to be in a hurry. He moves along on the ground very very slowly. He has a lazy walk. He is not afraid of most animals in the forest. His sharp quills keep him safe.

The quills grow in the porcupine's skin. Most of the time, his quills stay down on his back. But if an animal tries to hurt him, he gets mad. He makes his quills stand up.

Sometimes a young bobcat will try to catch a porcupine. She will try to bite him. If this happens, the porcupine turns away from the bobcat. He turns his back. Then he swings his tail.

The quills are held in the porcupine's skin. But they are not held tight. When he swings his tail, he hits the bobcat in the face. Some of the quills come out of his tail. They go into the bobcat's skin.

The sharp quills hurt the bobcat very much. There are tiny hooks at the end of each quill. The tiny hooks stick into the bobcat's skin. When she tries to pull out the quills, the little hooks will rip her skin.

Some people say that a porcupine can shoot his quills. But this is not true. Some quills come out when he swings his tail. Some quills fall out if he shakes himself. But he can not shoot his quills into the air.

The porcupine lives in a hole under a rock. He stays under the rock in the day. At night he comes out.

He eats vegetables. He hunts for vegetables in the night. In the summer, he eats clover and green leaves. He eats seeds and fruit. And he just loves to eat apples. He loves apples best of all.

In the winter, he can not find green vegetables and fruit. So he eats tree bark. The porcupine is a very good tree climber. He has sharp thick toenails on his feet. His sharp toenails help him to climb trees.

The porcupine has sharp front teeth too. He bites off the tree bark with his sharp front teeth. His sharp front teeth are very strong.

Sometimes the porcupine walks out of the forest. He walks into a farmer's barn. He bites on things that have a salty taste. He bites on wood and ax handles.

The porcupine loves the taste of salt. People have the taste of salt on their hands. The porcupine bites on things that people have touched with their salty hands.

If you live on a farm, maybe a porcupine will surprise you. Maybe a porcupine will come to your barn late at night. Maybe he will bite salty wood with his sharp front teeth. Watch for him!

The Chipmunk

It is fall. The little chipmunk chatters in the forest. She runs across a rock. She scampers in the leaves. She runs around a tree. She looks for acorns.

When the chipmunk finds an acorn, she stands up. She stands up on her back legs. She holds the acorn in her two front paws. She nibbles on the acorn with her big front teeth.

The chipmunk's big front teeth are very sharp. They have a yellow-orange color. There are two big teeth on the top and two big teeth on the bottom. The chipmunk bites acorns and nuts with her four sharp yellow-orange teeth.

After the chipmunk eats her dinner, she washes her face. She licks her front paws. She rubs her wet paws over her face. She likes to keep clean.

As soon as she feels clean, she scampers away. She hunts for more food. The chipmunk eats many kinds of food. She eats nuts and acorns. She eats fruit and seeds. She eats small bird eggs. She eats grasshoppers too.

The chipmunk needs to find a lot of food. She has to eat a big dinner. She also has to store up food for the winter.

The chipmunk lives under the ground. She lives in a house that is called a burrow. The chipmunk makes her own burrow. She digs a hole in the ground. She makes the hole go down deep. At the end of the hole, she makes a big room.

The chipmunk keeps her winter food in the big room. She brings food to the big room in the fall. She picks up food in her mouth and carries it to her burrow.

34

The chipmunk has a very special mouth. She has special cheeks in her mouth. The skin in her cheeks can stretch. She can hold a lot of nuts and seeds in her big cheeks. She can carry food to her burrow under the ground.

The chipmunk brings dry leaves into her burrow too. She rolls up one dry leaf at a time. She holds the dry leaf in her mouth. Then she carries it into her burrow. She covers up her food with all her dry leaves. She sleeps on her dry leaves in the winter.

The chipmunk is quiet in the winter. She sleeps most of the time. She rolls into a furry ball. She hardly moves.

After the chipmunk sleeps for a week, she will wake up. She will eat some of the food in her burrow. She will eat a few seeds and a few nuts. Then she will go back to sleep. She will do this all winter long.

Go to the forest in the summer or the fall. Sit down. Stay very quiet. You may hear the chipmunks chattering. You may see them running and scampering all over the ground.

The Woodpecker

Rat-a-tat-tat, rat-a-tat-tat. You can hear the woodpecker in the woods. You can hear the woodpecker banging on the trees. You can hear the woodpecker drilling holes. Rat-a-tat-tat, rat-a-tat-tat.

The woodpecker has a very strong beak. His beak is a little bit flat at the end. It is as sharp as a knife. He cuts into the wood with his sharp beak. He drills holes in the trees.

He drills holes so he can find food. Insects hide under the tree bark. The woodpecker drills in the tree bark so he can catch all the insects. He drills for his dinner.

The woodpecker gets the insects with his long tongue. His tongue is longer than his beak. It is sticky. He can lick insects out of the tree bark with his long sticky tongue.

The tip of his tongue has tiny hooks on it. The tiny hooks are pointed. The woodpecker puts his tongue under the tree bark or in a small hole. He pokes insects with the tiny hooks on the tip of his tongue. The insects stick to his tongue. Then he pulls them out and he eats them.

The woodpecker holds on to the tree bark with his sharp claws. He has to hold on tight. His tail is a big help to him. The feathers in his tail are very stiff. They stick out in back. They rest against the tree trunk.

The woodpecker holds on with his sharp claws. He drills with his sharp beak. And his stiff tail feathers help him to stay in one place.

The woodpecker can drill and drill with his sharp beak. But his head will never get hurt. He has very strong bones in his head. He has a very strong beak.

The woodpecker can drill and drill. But wood chips never get into his nose. He has two little nose holes in his beak. Tiny feathers cover up his nose holes. Air can get into his nose holes. Wood chips can not get in.

When the woodpecker is ready to have a family, he drills a big hole in the tree trunk. He drills and drills with his sharp beak. Chips of bark and wood fly into the air. The hole gets bigger and bigger. The hole will become a nest.

Soon there will be baby woodpeckers in the woods. When the babies grow up, they will hang on the tree bark. They will hunt for insects. They will drill holes in the trees. Rat-a-tat-tat, rat-a-tat-tat.

The Opossum

The opossum is a strange animal. She is about as big as a cat. She has pointed teeth like a dog. She has a long skinny tail like a rat. And she has a round pink nose like a pig.

The opossum lives in a hole in a tree. She is a very good climber. Her four paws work like hands. Her tail works like a hand too.

The opossum has five pink toes on each paw. Her toes are spread out. The claws on her toes are very sharp. Her claws can dig into the tree bark. Her toes can wrap around a thin branch.

Her long skinny tail has no fur. When the opossum is little, she can hang by her tail. When she gets bigger, she uses her tail for climbing. She holds on to branches with her tail. She carries things with her tail.

The opossum hunts for food in the night. She eats many kinds of food. She likes to eat mice and frogs. She likes to nibble on mushrooms. She likes to eat eggs and fruit and nuts.

The opossum can not run very fast. If a dog tries to hurt her, she backs up. She growls and she shows her teeth. If the dog won't go away, the opossum falls on the ground. She rolls over. She looks like she is dead.

When the opossum is playing dead, no one can make her move. The dog can bite her. He can poke her with his paw. But she still will not move at all. Soon the dog will go away. He thinks the opossum is really dead.

When the opossum is safe, she will get up. She will run back home. She will climb up her tree and hide in her tree hole.

The opossum will have babies in the spring. Before her babies are born, she makes a soft nest. She looks for leaves. She picks up a lot of leaves with her tail. Then she carries them home.

A mother opossum can have as many as twenty babies. Each baby will be very very tiny. Each baby will be about as big as a bee. Twenty babies could fit inside of one teaspoon!

The babies don't look like opossums when they are born. They have no fur. They have no eyes. They have two strong legs in front. But their back legs look like two little bumps.

The mother opossum has a pouch on her belly. Her pouch is like a pocket. There is fur inside of her pouch. There are thirteen tiny nipples in her pouch. Milk comes out of the nipples.

The tiny baby opossums need to get to their mother's pouch. They need to stay warm. They need to drink milk so they can stay alive.

It is hard for the tiny babies to find their mother's pouch. They have to work. They have to crawl up their mother's fur. They have tiny claws on their two front feet. They have to hold on tight with their tiny claws. They have to pull on their mother's fur. They have to climb up her belly before they can find her pouch.

When the babies find the pouch, they crawl inside. Then each baby tries to find a nipple. There may be twenty babies. But there are only thirteen nipples. If a baby can not find a nipple, it will not live. Some babies die.

Each baby holds on to a nipple with its mouth. It holds on tight. The babies drink a lot of milk. They grow in their mother's pouch. They stay in her pouch for two months.

When the babies come out of the pouch, they look like real opossums. They have four legs and four feet. They have two shiny black eyes. They have lots and lots of gray and white fur.

Now the furry little opossums get to ride on their mother's back. They hold on to her fur with their little feet. They wrap their little tails around their mother's big tail. Sometimes they wrap their tails around their mother's nose.

When the babies get bigger, they will leave their mother. They will find their own food. In the fall, all the opossums will eat a lot. They will get very fat. When winter comes, each opossum will find a hole under the ground. The opossums will sleep under the ground most of the winter.

Sometimes an opossum will wake up in the winter. She will go out of her hole. She will run across the snow. Her tracks will look like little stars.

This winter, take a walk in the forest. Maybe you will see little star-tracks in the snow.

It is raining outside. The little red
newt runs across the forest floor.
He runs across the rocks. He runs
across the bright green moss.

The little red newt likes to run in
the rain. He likes to get wet. He
hunts for food on rainy days. He
hunts at night too.

On sunny days, the red newt
hides under logs and rocks. He will
not hunt. The sun is too hot for
him. He has to keep his skin wet. If
his skin gets too dry, he will die. So
he stays near the wet ground.

The red newt has a long sticky
tongue. He can make his tongue
move in and out of his mouth very
quickly. He can catch insects with
his sticky tongue.

The red newt eats lots and lots of
insects. He grows. His skin gets
very tight. The skin on his head
pops open. When this happens, the
red newt wiggles out. He wiggles
out of his old skin.

As soon as he wiggles out of his
old skin, he eats it all up. Nothing is
left of his old skin.

The Red Newt

Now the newt's skin is nice and fresh. It has a bright red-orange color. It has a row of black and red freckle-spots on each side.

The red newt did not always live in the forest. He did not always have red skin. He was born in a pond. He came out of a little egg. He was a tadpole. He was colored yellow-green.

The tadpole came out of his egg in the spring. He stayed in the pond all summer long. At first, he had two legs in back and no legs in front. His front legs were just two little bumps. He had a flat tail.

The tadpole's head was big. There were gills on each side of his head. The gills looked like a group of small feathers.

The tadpole used his gills to get oxygen. He got oxygen from the water. He found food in the water too. He ate worms and insects that lived in the pond.

When the tadpole got bigger, his front legs started to grow. At the end of the summer, his gills got smaller. His flat tail got thinner. His yellow skin turned red.

In the fall, he left the pond. He went to the forest. He was not a tadpole any more. He did not have gills. He had four legs. He had a thin tail. He was a red newt.

The red newt started his life in the pond. Then he moved to the forest. But he will only stay in the forest for one or two years. He will sleep in the ground in the winter. He will run around in the spring and the summer. Then one fall, he will leave. He will leave the forest.

When the red newt leaves, he will walk back to the water. He will grow new gills. His tail will grow to be long and wide. His bright red skin will turn green. And he will spend the rest of his life swimming in the pond.

The red newt has a strange life. First he is born in the water. Then he lives on land for one or two years. Then he returns to the water.

If you go to the forest in the summer, you may find a little red newt. On a sunny day, you will have to peek under a rock or a log. But on a rainy day, watch out! The little red newt may run right across your path.

The Little Brown Bat

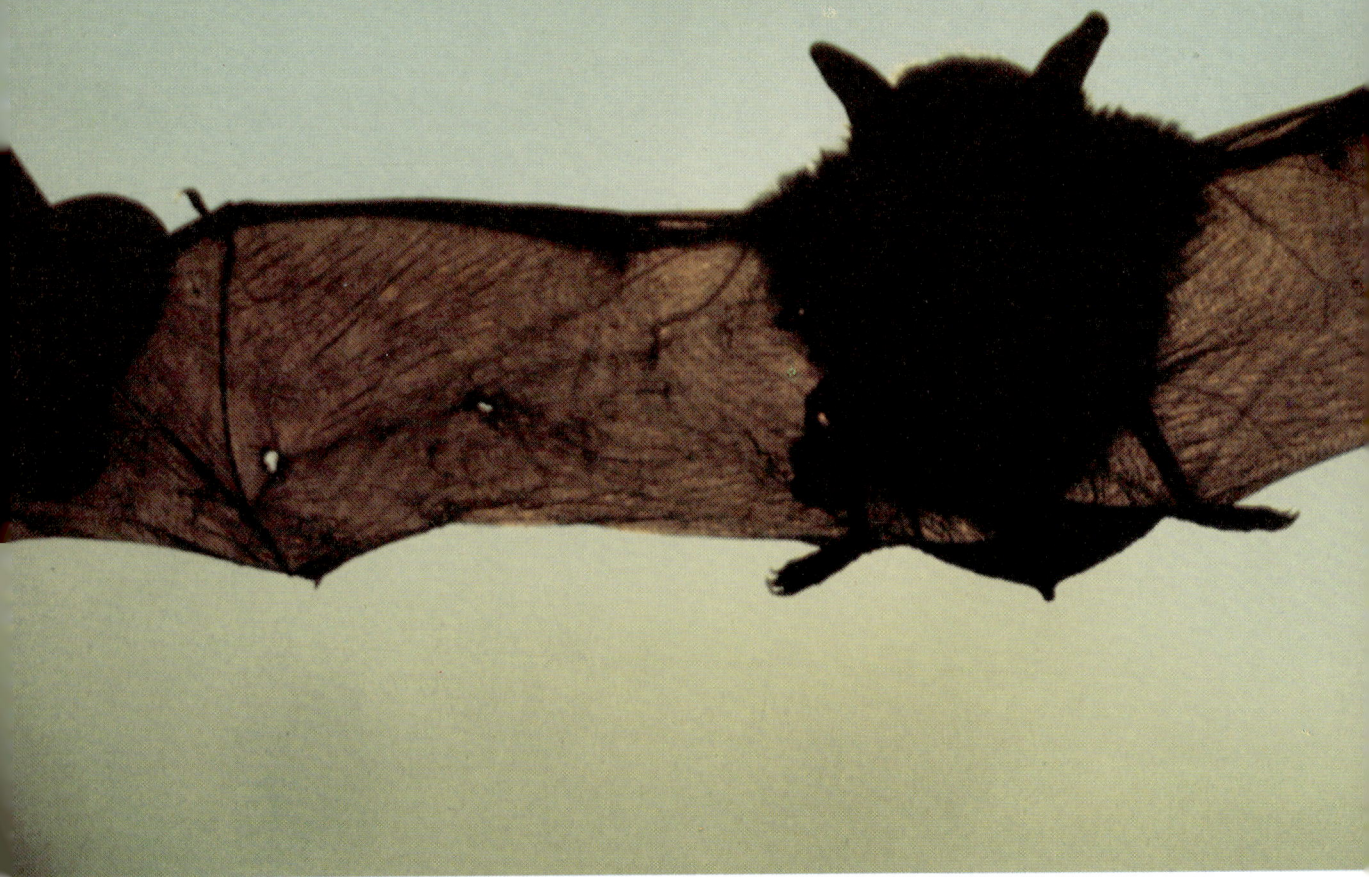

A little brown bat looks like a mouse. But she is not a mouse. A little brown bat flies like a bird. But she is not a bird.

The little brown bat is a mammal. All bats are mammals. Bats are the only mammals in the world that can fly.

The little brown bat has soft brown fur on her body. She has five fingers on each hand. The bones in her fingers are very very long. These long finger bones are inside of her wings. They are covered up with a thin sheet of skin. This sheet of skin is very thin. It feels like rubber.

The bat has no fur on her thin wings. If you look at a bat's wings up close, you can see her arm bones and her long finger bones inside. You can also see her thumbs.

Her thumbs are not covered by the thin sheet of skin. They are out in the open. Her two thumbs are not long. They are short. Each thumb has a claw. The bat can hold on to a tree trunk with one of her thumb-claws. She can clean her ears with her thumb-claws too.

The bat has two legs and a tail. The thin sheet of skin that covers her wings also covers her legs and her tail. When the bat flies, she moves her hand-wings up and down. She moves her legs up and down too.

The thin sheet of skin covers up the bat's legs. But it stops at her feet. Her two little feet are not covered by the thin sheet. They are free.

The bat has five little toes on each foot. Each toe has a sharp claw. These sharp claws are very strong.

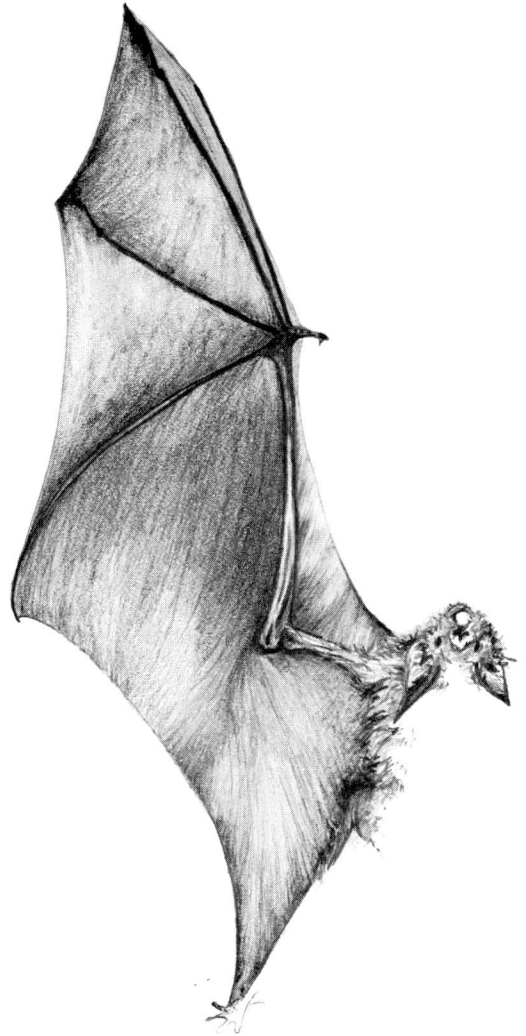

When she is not flying, the little brown bat hangs upside down in her tree or in her cave. She holds on with her sharp claws. A lot of other little brown bats hang upside down with her. They all sleep upside down.

At night, the little brown bats wake up. They wake up so they can go hunting. They hunt for insects. All little brown bats like to eat insects.

Sometimes the little brown bat can catch insects in her mouth when she is flying. Sometimes she hits insects with her tail skin and grabs them with her mouth. Other times she hits insects with her wings.

The little brown bat has thirty-eight tiny teeth. She bites the insects with her tiny teeth. Then she swallows them. She can eat over a thousand tiny insects in just one night.

When the little brown bat is hunting, she flies in the forest. She flies in a zigzag way. She flies in and out of all the trees. She flies fast. She turns fast. It is night. It is dark outside. But the little brown bat never hits a tree or a branch.

The little brown bat has two small black eyes. But she can not see in the dark with her eyes. So she has a special way of moving around in the dark.

When the little brown bat flies in the forest, she cries out. Her cries have a very very high sound. Her cries are so high that you can not hear them with your ears. But the brown bat can hear them. She can hear sounds that you can not hear.

The little brown bat cries out. The sounds of her high cries bounce off of the trees. They bounce off of other animals. The high sounds bounce back to the bat. She hears the bouncing sounds. She hears the bouncing sounds with her good ears.

When the bat hears a bouncing sound, she knows that something is in her way. She turns quickly. She will not hit a tree. And if you are standing in the forest, she will not hit you. Her high cries will bounce off of your body. She will know that you are there.

The little brown bat is a very good mother. She has just one baby a year. She takes good care of her baby.

When she is ready to have a baby, she hangs by her thumbs. She curls up her tail skin to make a basket. When the baby comes out of the mother bat, it falls into the basket. If the mother did not make a basket, the baby would fall on the ground.

After the baby is born, he hangs on to his mother's fur. The baby has sharp little claws on his toes. He hangs on to his mother's fur with his sharp claws. He drinks milk from his mother too.

At night, the mother bat will go hunting. She takes her young baby with her. The baby hangs on to his mother's fur. The mother flies around and catches insects.

When the baby gets bigger, the mother bat can not carry him around. So she hangs him up. She hangs him up in a tree or in a cave. The baby hangs upside down. The mother hunts by herself.

Many people do not like bats. Many people are afraid of bats. But bats are not really so bad. They eat lots and lots of insects in the forest. They eat insects that can hurt trees. The bats are good for the forest. Be glad they are there.

CREDITS

Designed by Gale William Ikola and Cyril John Schlosser
Illustrated by Maarja Roth-Evenson

Photo Credits

Craig Blacklock: 46
Margaret B. Brandow/Tom Stack & Associates: 48
Perry Covington/Tom Stack & Associates: 35
Ron Dillon/Tom Stack & Associates: 25
Warren Garst/Tom Stack & Associates: 45
Zig Leszczynski/Animals Animals Enterprises: 16, 40, 49
Glenn Maxham, Maxham Films, Inc.: 14-15, 21, 22-23, 27, 50-51
L. David Mech: 19, 31, 41
J. Donald Meyer: 36
Minnesota Department of Economic Development: 28
David Mork: 54
Tom and Ceil Ramsey, TCR Productions: 9
Cyril A. Reilly: 34
Lynn L. Rogers: cover, 18, 26, 32-33
Lynn M. Stone: 6-7, 24, 37, 39 (both), 53
Marty Stouffer/Animals Animals Enterprises: 42
Jan L. Wassink/Tom Stack & Associates: 10
Ernest Wilkinson/Animals Animals Enterprises: 12-13
C. M. Wright/Tom Stack & Associates: 28-29

56

INDEX